The Open University

M3

C2

Compactness

This publication forms part of an Open University course. Details of this and other Open University courses can be obtained from the Student Registration and Enquiry Service, The Open University, PO Box 197, Milton Keynes, MK7 6BJ, United Kingdom: tel. +44 (0)870 333 4340, e-mail general-enquiries@open.ac.uk

Alternatively, you may visit the Open University website at http://www.open.ac.uk where you can learn more about the wide range of courses and packs offered at all levels by The Open University.

To purchase a selection of Open University course materials, visit the webshop at www.ouw.co.uk, or contact Open University Worldwide, Michael Young Building, Walton Hall, Milton Keynes, MK7 6AA, United Kingdom, for a brochure: tel. +44 (0)1908 858785, fax +44 (0)1908 858787, e-mail ouwenq@open.ac.uk

The Open University, Walton Hall, Milton Keynes, MK7 6AA.

First published 2006.

Copyright © 2006 The Open University

All rights reserved; no part of this publication may be reproduced, stored in a retrieval system, transmitted or utilised in any form or by any means, electronic, mechanical, photocopying, recording or otherwise, without written permission from the publisher or a licence from the Copyright Licensing Agency Ltd. Details of such licences (for reprographic reproduction) may be obtained from the Copyright Licensing Agency Ltd, 90 Tottenham Court Road, London W1T 4LP.

Open University course materials may also be made available in electronic formats for use by students of the University. All rights, including copyright and related rights and database rights, in electronic course materials and their contents are owned by or licensed to The Open University, or otherwise used by The Open University as permitted by applicable law.

In using electronic course materials and their contents you agree that your use will be solely for the purposes of following an Open University course of study or otherwise as licensed by The Open University or its assigns.

Except as permitted above you undertake not to copy, store in any medium (including electronic storage or use in a website), distribute, transmit or re-transmit, broadcast, modify or show in public such electronic materials in whole or in part without the prior written consent of The Open University or in accordance with the Copyright, Designs and Patents Act 1988.

Edited, designed and typeset by The Open University, using the Open University TEX System.

Printed and bound in the United Kingdom by The Charlesworth Group, Wakefield.

ISBN 0 7492 4135 7

Contents

Introduction		**4**
	Study guide	4
1	**Introducing compactness**	**5**
	1.1 Covers and the unit interval	5
	1.2 Open covers and the Extreme Value Theorem	8
2	**Compact topological spaces**	**11**
	2.1 Definition of compactness	11
	2.2 General results about compact spaces	13
	2.3 Products of compact topological spaces	15
3	**Compactness in Hausdorff spaces**	**18**
	3.1 Hausdorff spaces	18
	3.2 Compact sets in Hausdorff spaces	21
4	**Compactness in Euclidean spaces**	**26**
	4.1 Compact subsets of Euclidean spaces	26
	4.2 The Extreme Value Theorem	27
5	**Surfaces and compactness**	**29**
Solutions to problems		**30**
Index		**32**

Introduction

In *Unit A1*, we observed that if $f\colon [a,b] \to \mathbb{R}$ is a continuous function, then the Extreme Value Theorem holds — f is bounded and attains its bounds: that is, as Figure 0.1 illustrates, there are $c, d \in [a, b]$ such that

$$f(c) \leq f(x) \leq f(d) \quad \text{for all } x \in [a, b].$$

We also stated that such a result need not be true for continuous functions on intervals such as (a, b) and $[a, \infty)$. In this unit, we investigate what property of $[a, b]$ guarantees the boundedness of continuous functions on it. One of our motivations for doing this is to try to find a general version of the Extreme Value Theorem that holds for continuous functions from *any* topological space to $\mathbb{R}$.

The notion we introduce is one that we discussed briefly in Block B, that of *compactness*. We shall show that the interval $[0, 1]$ is compact whereas the intervals $(0, 1)$ and $[0, \infty)$ are not, and it is this property that ensures that continuous functions on $[0, 1]$ are bounded and attain their extreme values. We shall see that the results generalize to the interval $[a, b]$.

One important problem that we consider in this unit is how to determine whether a given subset of a Euclidean space is compact: this question turns out to have a surprisingly simple answer. This plays a key role in our proof of a generalized Extreme Value Theorem.

Unit A1, Theorem 2.5.

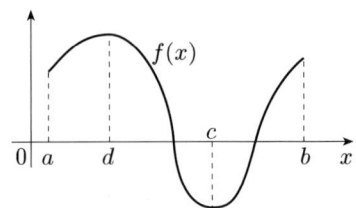

Figure 0.1

Unit B1, Subsection 1.5.

Study guide

Although this unit is not particularly long, the ideas introduced within it are subtle, and you may find that it takes some effort and practice before you are fully comfortable with them. This is normal — the material is often regarded as quite difficult.

The most important ideas of this unit appear in Sections 2 and 3. You will probably need to spend most of your time studying these two sections.

In Section 1, *Introducing compactness*, we introduce the notion of a *cover*, and start to investigate how covers can be used to encapsulate various properties of continuous functions on the closed bounded interval $[0, 1]$.

In Section 2, *Compact topological spaces*, we give a formal definition of compactness and discuss some basic properties of compact sets.

In Section 3, *Compactness in Hausdorff spaces*, we restrict our attention to a particular class of topological spaces that includes topologies defined by metrics — the *Hausdorff spaces*. We show that, in a Hausdorff space, compact sets have some extra properties that they need not possess in an arbitrary topological space.

In Section 4, *Compactness in Euclidean spaces*, we derive a simple characterization of compact sets in Euclidean spaces. This enables us to prove a generalization of the Extreme Value Theorem. It is important that you understand how to use these results.

Finally, in Section 5, *Surfaces and compactness*, we relate the notion of compactness described in Block B to the one discussed in this unit. This is a very short section and is not assessed.

There is no software associated with this unit.

1 Introducing compactness

> After working through this section, you should be able to:
> ▶ give examples of *open covers* of sets in topological spaces;
> ▶ decide whether a given open cover of an interval has a *finite subcover*.

Let $f: [0,1] \to \mathbb{R}$ be continuous (for the Euclidean topologies). Then the Extreme Value Theorem tells us that there are $c, d \in [0,1]$ such that

$$f(c) \leq f(x) \leq f(d) \quad \text{for all } x \in [0,1].$$

Unit A1, Theorem 2.5.

Since this holds for any $f \in C[0,1]$, it must arise out of the interaction between properties of the interval $[0,1]$ with arbitrary continuous functions, rather than out of the detailed properties of any particular function.

Recall that $C[0,1]$ is the set of continuous functions from $[0,1]$ to $\mathbb{R}$.

Now consider $\mathbb{R}^2$, let $B = B_{d^{(2)}}[\mathbf{0}, 1]$ be the *closed* unit ball in $\mathbb{R}^2$ and let $g: B \to \mathbb{R}$ be a continuous function (again for the Euclidean topologies). Are there points $\mathbf{c}, \mathbf{d} \in B$ such that

$$g(\mathbf{c}) \leq g(\mathbf{x}) \leq g(\mathbf{d}) \quad \text{for all } \mathbf{x} \in B?$$

What happens if, instead of B, g is defined on the unit *sphere* in $\mathbb{R}^2$, or on the *open* unit ball? What happens if g is defined on some other, more abstract, topological space?

In order to find an answer to these questions, we investigate what topological property of $[0,1]$ gives rise to the fact that functions in $C[0,1]$ are bounded and attain their bounds. Since our goal is to find a property that makes sense in more general topological spaces, we use open sets to try to understand the special properties of $[0,1]$.

1.1 Covers and the unit interval

We begin our investigations by studying collections of open sets that *cover* $[0,1]$. You will see later in this section that such collections hold the key to understanding exactly what it is about the interval $[0,1]$ that enables the Extreme Value Theorem to hold.

Definition

Let $(X, \mathcal{T})$ be a topological space, and let $A \subseteq X$.

A collection $\mathcal{S}$ of subsets of X is a **cover** of A if

$$A \subseteq \bigcup_{U \in \mathcal{S}} U;$$

that is, for each $a \in A$, there exists $U \in \mathcal{S}$ such that $a \in U$.

$\mathcal{S}$ is an **open cover** of A if, in addition, $\mathcal{S} \subseteq \mathcal{T}$.

Thus an open cover of A is a cover of A that consists entirely of open sets.

Remark

If we wish to emphasize the topology, we say $\mathcal{T}$-open cover.

There are many different open covers of $[0,1]$. We begin with the following example.

Worked problem 1.1

Show that
$$S_1 = \{(x - \tfrac{1}{10}, x + \tfrac{1}{10}) : x \in [0,1]\}$$
is an open cover of $[0,1]$ for the Euclidean topology on $\mathbb{R}$.

Solution

Let $U_x = (x - \tfrac{1}{10}, x + \tfrac{1}{10})$. If $x \in [0,1]$, then $x \in U_x$ and so
$$[0,1] \subseteq \bigcup_{x \in [0,1]} U_x = \bigcup_{U_x \in S_1} U_x.$$

Also, each set U_x in S_1 is open. So S_1 is an open cover of $[0,1]$. ∎

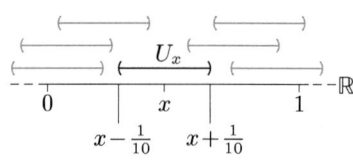

Figure 1.1

You may have noticed that many of the sets in S_1 overlap, suggesting that $[0,1]$ can be covered by a smaller collection of sets. In fact $[0,1]$ can be covered by just *finitely* many of the sets in S_1.

Problem 1.1

Find a finite subcollection of sets in S_1 that covers $[0,1]$.

Now consider
$$S = \{(0, q) : q \in \mathbb{Q} \cap (0, \infty)\}.$$

This is a collection of open sets with respect to the Euclidean topology on $\mathbb{R}$, but it is not a cover of $[0,1]$, for no set in S contains the point 0. However,
$$S_2 = S \cup \{(-\tfrac{1}{100}, \tfrac{1}{100})\}$$
is an open cover of $[0,1]$.

Problem 1.2

(a) Show that S_2 is an open cover of $[0,1]$.
(b) Find a finite subcollection of sets in S_2 that covers $[0,1]$.

We shall show that *any* open cover of $[0,1]$ with respect to the Euclidean topology on $\mathbb{R}$ contains a finite subcollection of sets that covers $[0,1]$. It is this property of $[0,1]$ that enables the Extreme Value Theorem to hold. Before proving this result, we need the following definition.

Definition

Let $(X, \mathcal{T})$ be a topological space, let $A \subseteq X$ and let S be a cover of A. A collection $\mathcal{R}$ of subsets of X is a **finite subcover** of A from S if:

(a) $\mathcal{R} \subseteq S$;
(b) $A \subseteq \bigcup_{U \in \mathcal{R}} U$;
(c) $\mathcal{R}$ is a finite collection of sets.

Remark

This definition does not require the sets in $\mathcal{R}$ to be open. However, in most of our applications, the starting collection $\mathcal{S}$ is an open cover, and so any finite subcover from $\mathcal{S}$ automatically consists of open sets.

Theorem 1.1

With respect to the Euclidean topology on $\mathbb{R}$, every open cover of the interval $[0, 1]$ has a finite subcover.

Proof Let $\mathcal{S}$ be an open cover of $[0, 1]$. We must show that $\mathcal{S}$ contains a finite subcover of $[0, 1]$.

Let A be the set of points $x \in [0, 1]$ for which the subinterval $[0, x]$ is covered by a finite subcollection of sets from $\mathcal{S}$.

> Considering A is not an obvious thing to do, but it lies at the heart of the proof.

Certainly $0 \in A$, since 0 belongs to at least one set from $\mathcal{S}$, and that set (alone) is a finite subcover of 0. Hence A is not empty.

Observe that if $x \in A$ and $0 \leq y \leq x$, then $y \in A$. For, if $\{U_1, U_2, \ldots, U_n\}$ is a finite subset of $\mathcal{S}$ that covers $[0, x]$, then it also covers $[0, y] \subseteq [0, x]$. Thus A is an interval.

The set A is bounded above by 1, so it must have a least upper bound; denote it by a. Clearly $0 \leq a \leq 1$. If we can show that

$$a \in A \quad \text{and} \quad a = 1,$$

then the proof will be complete.

> The existence of the least upper bound follows from the fact that we are working in $\mathbb{R}$.

Proof that $a \in A$

If $a = 0$, then we know that $a \in A$. So suppose that $a \in (0, 1]$. There is a set $U \in \mathcal{S}$ with $a \in U$. Since U is open, we can find r with $0 < r < a$ such that $(a - r, a + r) \subseteq U$. Consider now the point $b = a - \frac{1}{2}r \in (0, 1]$. This point certainly lies in A — for otherwise each point x between b and a is not in A, contradicting the fact that a is the *least* upper bound of A. Thus the interval $[0, b]$ has a finite subcover $\{U_1, U_2, \ldots, U_n\}$ from $\mathcal{S}$. But then $\{U\} \cup \{U_1, U_2, \ldots, U_n\}$ is a finite subcover from $\mathcal{S}$ of $[0, a]$. It follows that $a \in A$.

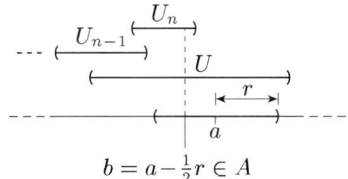

Figure 1.2

Proof that $a = 1$

Suppose, for a contradiction, that $a < 1$. Then, since $a \in A$, the interval $[0, a]$ has a finite subcover $\{U_1, U_2, \ldots, U_n\}$ from $\mathcal{S}$. In particular, $a \in U_i$, for some $i \in \{1, 2, \ldots, n\}$. For this open set U_i, we can find $0 < r < 1 - a$ such that $(a - r, a + r) \subseteq U_i$. Hence $a + \frac{1}{2}r \in U_i$, and the interval $[0, a + \frac{1}{2}r]$ has a finite subcover from $\mathcal{S}$ — namely, $\{U_1, U_2, \ldots, U_n\}$. Thus since $a + \frac{1}{2}r < a + r < 1$, it follows that $a + \frac{1}{2}r \in A$, contradicting the fact that a is an *upper bound* of A. We deduce that $a = 1$.

It follows that $\mathcal{S}$ contains a finite subcover of $[0, 1]$. ∎

Remark

This proof is a classic, created even before formal set theory was invented. An analogous proof can be used to show that any closed bounded interval $[a, b]$ has this property.

We conclude this subsection by noting that it is important that we consider *open* covers of $[0, 1]$ — if the cover is not open, there may be no finite subcover.

Problem 1.3

With respect to the Euclidean topology on $\mathbb{R}$, show that there is an infinite cover of $[0, 1]$ by *non-open* sets that has no finite subcover.

1.2 Open covers and the Extreme Value Theorem

Recall that we are trying to identify some property involving open sets that explains why the Extreme Value Theorem holds for the interval $[0, 1]$ but not for the intervals $(0, 1)$ and $[0, \infty)$. We have seen that every open cover of $[0, 1]$ has a finite subcover. We now show that the intervals $(0, 1)$ and $[0, \infty)$ do not have this property.

Worked problem 1.2

Let $\mathbb{R}$ have the Euclidean topology and let A be the bounded open set $(0, 1)$. Find an open cover of A that has no finite subcover.

Solution

There are many open covers with this property. One such open cover is

$$\mathcal{S} = \{(\tfrac{1}{n}, 1) : n = 2, 3, 4, \ldots\} = \{(\tfrac{1}{2}, 1), (\tfrac{1}{3}, 1), (\tfrac{1}{4}, 1), \ldots\}.$$

For each $x \in (0, 1)$, the interval $(\tfrac{1}{n}, 1)$ contains x if $n \in \mathbb{N}$ is chosen so that $\tfrac{1}{n} < x$. Thus $\mathcal{S}$ is a cover of $(0, 1)$. Also, the intervals in $\mathcal{S}$ are open. So $\mathcal{S}$ is an open cover of $(0, 1)$.

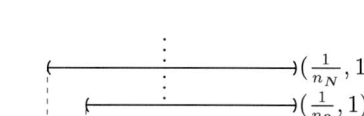
For example, if $x = \tfrac{1}{100}$, then $(\tfrac{1}{101}, 1)$ contains x.

Suppose that $\mathcal{R}$ is a finite subcollection of $\mathcal{S}$. We must show that $\mathcal{R}$ does not cover $(0, 1)$. Since $\mathcal{R}$ is finite, we can write

$$\mathcal{R} = \{(\tfrac{1}{n_1}, 1), (\tfrac{1}{n_2}, 1), \ldots, (\tfrac{1}{n_N}, 1)\},$$

where we may assume that $n_1 < n_2 < \cdots < n_N$. Then

$$\bigcup_{U \in \mathcal{R}} U = (\tfrac{1}{n_N}, 1),$$

and so no point $x \in (0, \tfrac{1}{n_N}]$ is covered. Thus $\mathcal{R}$ is not a cover of $(0, 1)$. ∎

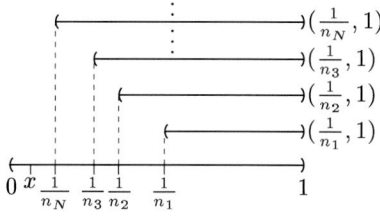
Figure 1.3

Problem 1.4

Let $\mathbb{R}$ have the Euclidean topology and let A be the closed unbounded set $[0, \infty)$. Find an open cover of A that has no finite subcover.

The results of Worked problem 1.2 and Problem 1.4 can be generalized to show that not all open covers of the open interval (a, b), the half-open intervals $(a, b]$, $[a, b)$, or the unbounded intervals $[a, \infty)$, (a, ∞), $(-\infty, a]$ and $(-\infty, a)$ have finite subcovers. Thus the finite subcover property applies only to intervals that are closed and bounded.

We now see that our statement that every open cover of $[0, 1]$ contains a finite subcover seems to capture some fundamental property of the closed bounded interval $[0, 1]$ that other types of interval fail to possess. Moreover, this property is phrased entirely in terms of open sets, and so can be used as the basis of a definition in a general topological space. We shall make such a definition in Section 2.

We end this section by showing how this property of the unit interval can be used to show that the Extreme Value Theorem holds for $[0,1]$ — that is, continuous functions from $[0,1]$ to $\mathbb{R}$ are bounded and attain their bounds.

We first show that real-valued continuous functions on $[0,1]$ are bounded.

Lemma 1.2

Let $f\colon [0,1] \to \mathbb{R}$ be continuous for the Euclidean topologies. Then f is bounded.

Proof Consider
$$\mathcal{S} = \{U_i : U_i = (i-1, i+1),\, i \in \mathbb{Z}\}.$$
Each point $x \in \mathbb{R}$ belongs to an interval U_i. Also, each interval U_i is open. So $\mathcal{S}$ is an open cover of $\mathbb{R}$.

Corresponding to each set U_i is the subset $f^{-1}(U_i)$ of the domain $[0,1]$. Since f is continuous on $[0,1]$, $f^{-1}(U_i)$ is an open set in $[0,1]$. For each $x \in [0,1]$, $f(x)$ must belong to at least one of the covering sets U_i. So
$$\mathcal{R} = \{f^{-1}(U_i) : i \in \mathbb{Z}\}$$
is an open cover of $[0,1]$ — that is,
$$[0,1] \subseteq \bigcup_{i \in \mathbb{Z}} f^{-1}((i-1, i+1)).$$
Now Theorem 1.1 implies that we can find a finite subcover of $[0,1]$ from $\mathcal{R}$ — that is, we can find $N \in \mathbb{N}$ such that
$$[0,1] \subseteq \bigcup_{i=-N}^{N} f^{-1}(U_i).$$
Hence
$$f([0,1]) \subseteq \bigcup_{i=-N}^{N} U_i = (-N-1, N+1).$$
But this means that
$$-N-1 < f(x) < N+1 \quad \text{for all } x \in [0,1].$$
So f is bounded. ∎

Thus the finite subcover property of the unit interval is all that is needed to show that a continuous function on that interval is bounded. We now show that such a function attains its least upper bound.

Lemma 1.3

Let $f\colon [0,1] \to \mathbb{R}$ be continuous for the Euclidean topologies. Then f attains its least upper bound.

Proof By Lemma 1.2, f is bounded. Thus it has a least upper bound M: that is, $f(x) \leq M$ for all $x \in [0,1]$.

In order to show that f attains the value M, we give a proof by contradiction.

Assume that $f(x) < M$ for all $x \in [0,1]$. Then, for each $x \in [0,1]$, we can find $n \in \mathbb{N}$ for which $f(x) < M - \frac{1}{n}$, and so

$$f([0,1]) \subseteq \bigcup_{n \in \mathbb{N}} (-\infty, M - \tfrac{1}{n}).$$

Hence

$$[0,1] \subseteq f^{-1}\left(\bigcup_{n \in \mathbb{N}} (-\infty, M - \tfrac{1}{n})\right) = \bigcup_{n \in \mathbb{N}} f^{-1}((-\infty, M - \tfrac{1}{n})). \qquad \text{Unit A3, Theorem 2.7.}$$

Since f is continuous, $f^{-1}((-\infty, M - \tfrac{1}{n}))$ is open, for each $n \in \mathbb{N}$, and so

$$\{f^{-1}((-\infty, M - \tfrac{1}{n})) : n \in \mathbb{N}\}$$

is an open cover of $[0,1]$. Theorem 1.1 now implies that there is $N \in \mathbb{N}$ such that

This is the finite subcover property.

$$[0,1] \subseteq \bigcup_{n=1}^{N} f^{-1}((-\infty, M - \tfrac{1}{n})).$$

Now

$$\bigcup_{n=1}^{N} f^{-1}\left((-\infty, M - \tfrac{1}{n})\right) = f^{-1}\left(\bigcup_{n=1}^{N} (-\infty, M - \tfrac{1}{n})\right) \qquad \text{Unit A3, Theorem 2.7.}$$
$$= f^{-1}\left((-\infty, M - \tfrac{1}{N})\right).$$

Therefore

$$f([0,1]) \subseteq f\left(\bigcup_{n=1}^{N} f^{-1}((-\infty, M - \tfrac{1}{n}))\right) = (-\infty, M - \tfrac{1}{N}).$$

But this contradicts the definition of M as the *least* upper bound of f.

We deduce that f attains its least upper bound. ∎

A similar argument to that in the proof of Lemma 1.3 shows that f also attains its greatest lower bound. Combining these results we have a proof of the Extreme Value Theorem.

In proving this result, we have only used the finite subcover property of $[0,1]$, so clearly this property of $[0,1]$ is important. In the following sections we study topological spaces with similar properties, enabling us to prove a generalized version of the Extreme Value Theorem in Section 4.

2 Compact topological spaces

After working through this section, you should be able to:
- explain what is meant by a *compact* topological space;
- decide whether a given topological space is compact;
- describe some properties of compact topological spaces.

In the previous section, we investigated what distinguishes the interval $[0,1]$ from intervals such as $(0,1)$ and $[0,\infty)$, and guarantees that the Extreme Value Theorem holds for $[0,1]$. We saw that each open cover of $[0,1]$ has a finite subcover, and that this is not the case for open intervals or for unbounded intervals. Moreover, this property of $[0,1]$ is sufficient to allow us to show that the Extreme Value Theorem holds for $[0,1]$.

In this section, we investigate some of the properties of general topological spaces for which any open cover has a finite subcover. These investigations form a first step towards proving the Extreme Value Theorem for such spaces.

The proof is in Section 4.

2.1 Definition of compactness

We introduce a special name for a topological space with the property that every open cover has a finite subcover — such a space is said to be *compact*.

Definition

A topological space $(X, \mathcal{T})$ is **compact** if each open cover of X contains a finite subcover of X.

A subset $A \subseteq X$ is **compact** if $(A, \mathcal{T}_A)$ is compact.

The link between this definition of compactness and that given in *Unit B1*, in the context of surfaces, is made in Section 5.

$\mathcal{T}_A$ is the subspace topology on A inherited from $\mathcal{T}$.

Remarks

(i) When proving that a subset $A \subseteq X$ is compact, it is sufficient to take an open cover of A by sets in $\mathcal{T}$, since, if $U \in \mathcal{T}$, then, by the definition of the subspace topology, $U \cap A \in \mathcal{T}_A$. Moreover, if $V \in \mathcal{T}_A$, then there is $U \in \mathcal{T}$ for which $V = U \cap A$. Hence, if an open cover of A by sets in $\mathcal{T}$ has a finite subcover, then the corresponding open cover in $\mathcal{T}_A$ also has a finite subcover, and vice versa.

(ii) It follows from Theorem 1.1 that the interval $[0,1]$, with its Euclidean subspace topology, is compact. A similar argument can be used to show that each closed and bounded interval $[a,b]$ is compact.

(iii) If we wish to emphasize the topology, we say that A is $\mathcal{T}$-*compact* or *compact for* $\mathcal{T}$.

(iv) Warning: when trying to prove that a given space X is compact, it is tempting, but *wrong*, to choose some *particular* open cover and demonstrate that it has a finite subcover. This *does not* prove that X is compact. We must show that *no matter what choice* of open cover we make, it always has a finite subcover. Thus a proof of compactness typically begins by considering an *arbitrary* open cover $\mathcal{S}$ of X.

$(0,1)$ and $[0,\infty)$ are not compact, nor is any interval that is not both closed and bounded.

Problem 2.1

Let $(X, \mathcal{T})$ be a topological space. Show that $\varnothing$ is a compact subset of X.

We now examine some of our standard topological spaces to see which are compact. The most important examples for us to examine for compactness are subsets of $\mathbb{R}^n$. However, this requires us to develop further the theory of compactness; so we defer discussion of subsets of $\mathbb{R}^n$ until Section 4, when we will see that there is a simple classification of compact subsets of Euclidean spaces. Instead, we begin with the discrete and indiscrete topologies.

Example 2.1

Consider the *indiscrete topology* on a set X. Then, for any set $A \subseteq X$, each open cover contains at most two distinct sets — $\varnothing$ and X. Hence every open cover is finite, and so every subset of X is compact. ∎

The indiscrete topology is $\{\varnothing, X\}$.

Example 2.2

For the *discrete topology* things are different. Suppose that X is an infinite set, and let A be an infinite subset of X. Then to show that A is not compact, it is enough to exhibit one open cover of A that has no finite subcover: the collection $\{\{a\} : a \in A\}$ is such a cover. ∎

Here every subset of X is open.

Problem 2.2

Let $(X, \mathcal{T})$ be a topological space. Show that each finite subset of X is compact.

Worked problem 2.1 *a-deleted-point topology*

Let a be an element of a set X. Show that X is compact for the a-deleted-point topology $\mathcal{T}_a$.

A set U is in $\mathcal{T}_a$ if either $U = X$ or $a \notin U$.

Solution

Let $\mathcal{S}$ be an open cover of X. Since $\mathcal{S}$ covers X, it must contain an open set containing a. But the only open set containing a is X itself, so every open cover of X must contain X — and so contains the one-set subcover $\{X\}$. Hence $(X, \mathcal{T}_a)$ is compact. ∎

Problem 2.3

Let $\mathcal{T}$ be the either-or topology on $[-1, 1]$ in which a set U is open if either $0 \notin U$ or $(-1, 1) \subseteq U$. Show that, for this topology, $[-1, 1]$ is compact, but $[-1, 0)$ is not.

You met this topology in *Unit A3*, Problem 3.5.

Although we cannot yet classify the compact subsets of $\mathbb{R}$, we can show that $\mathbb{R}$ itself is not compact.

Problem 2.4

Show that $\mathbb{R}$ is not compact for the Euclidean topology, by finding an open cover of $\mathbb{R}$ that contains no finite subcover.

Problem 2.5

Show that $\mathbb{R}^2$ is not compact for the Euclidean topology, by finding an open cover of $\mathbb{R}^2$ that contains no finite subcover.

2.2 General results about compact spaces

Suppose that $(X, \mathcal{T})$ is a compact topological space. Are there any subsets of X that are necessarily compact, apart from finite subsets and X? The next result tells us that a compact topological space may have many compact subsets.

In Problem 2.2 you saw that finite sets are always compact.

Theorem 2.1

Each closed subset of a compact set is compact.

Proof Let $(X, \mathcal{T})$ be a compact topological space, and suppose that $A \subseteq X$ is closed. We must show that A is compact — that is, each open cover of A has a finite subcover.

Let $\mathcal{S}$ be an arbitrary open cover of A. We must show that $\mathcal{S}$ contains a finite subcover. Now $A \subseteq \bigcup_{U \in \mathcal{S}} U$, and so

$$X \subseteq A^c \cup \bigcup_{U \in \mathcal{S}} U.$$

Here we add in A^c.

But A^c is an open set, since A is closed, and so

$$\mathcal{R} = \mathcal{S} \cup \{A^c\}$$

is an open cover of X. Since $(X, \mathcal{T})$ is compact, there is a finite subcover of X from $\mathcal{R}$ — call it $\mathcal{P}$. But then

$$\{U \in \mathcal{P} : U \neq A^c\}$$

Removing A^c brings us back to a subcover of $\mathcal{S}$.

is a finite subcover of A from $\mathcal{S}$. Since $\mathcal{S}$ is an arbitrary open cover of A, we deduce that A is compact. ∎

If we know that a topological space is compact, then we can use Theorem 2.1 to identify other compact spaces. For example, since $[0, 1]$ is a compact subset of $\mathbb{R}$ with the Euclidean topology, any closed subset of $[0, 1]$ is also compact.

Problem 2.6

Let $[0, 1]$ have the Euclidean subspace topology. Show that $A = \{0\} \cup \{\frac{1}{n} : n \in \mathbb{N}\}$ is a compact subset of $X = [0, 1]$.

Since the compactness property is expressed entirely in terms of open sets, it should be no surprise to learn that *compactness is a topological invariant*. To prove this, we show that the image of a compact set under a continuous function is compact. This result is a key part of our programme to generalize the Extreme Value Theorem.

Theorem 2.2

Let $(X, \mathcal{T}_X)$ and $(Y, \mathcal{T}_Y)$ be topological spaces, let X be compact and let $f : X \to Y$ be $(\mathcal{T}_X, \mathcal{T}_Y)$-continuous. Then $f(X)$ is compact.

Proof Let $\mathcal{S}$ be an arbitrary open cover of $f(X)$. We must show that $\mathcal{S}$ contains a finite subcover.

If $x \in X$, then $f(x) \in U$ for some $U \in \mathcal{S}$ and so $x \in f^{-1}(U)$. Since f is continuous, $f^{-1}(U)$ is an open set for each $U \in \mathcal{S}$ and so
$$\mathcal{R} = \{f^{-1}(U) : U \in \mathcal{S}\}$$
is an open cover of X. Since X is compact, $\mathcal{R}$ has a finite subcover $\{f^{-1}(U_1), f^{-1}(U_2), \ldots, f^{-1}(U_n)\}$, say. Hence
$$X = \bigcup_{i=1}^{n} f^{-1}(U_i) = f^{-1}\left(\bigcup_{i=1}^{n} U_i\right),$$
Unit A3, Theorem 2.7.

and so
$$f(X) \subseteq \bigcup_{i=1}^{n} U_i.$$

Thus $\{U_1, U_2, \ldots, U_n\}$ is a finite subcover of $f(X)$ by sets from $\mathcal{S}$, and hence $f(X)$ is compact. ∎

Since a homeomorphism and its inverse function are both onto and continuous, it follows that

> if $(X, \mathcal{T}_X)$ and $(Y, \mathcal{T}_Y)$ are homeomorphic topological spaces, and one of them is compact, then so is the other.

We can restate this result as follows.

Corollary 2.3

Compactness is a topological invariant.

Remark

It follows that, if $(X, \mathcal{T}_X)$ and $(Y, \mathcal{T}_Y)$ are homeomorphic topological spaces, and one of them is not compact, then neither is the other.

Example 2.3

We have shown directly that the interval $[0, 1]$ is compact. Since $[a, b]$ is homeomorphic to $[0, 1]$, it follows that any closed and bounded interval $[a, b]$ is compact. Similarly, we have shown that $(0, 1)$ is not compact. Since (a, b) is homeomorphic to $(0, 1)$, it follows that no open interval (a, b) is compact. ∎

Problem 2.7

Give an example to show that the image of a compact set under a *discontinuous* function need not be compact.

Our final result in this subsection is a straightforward consequence of Theorem 2.2.

Corollary 2.4

Let $(X, \mathcal{T}_1)$ be a compact topological space and let $\mathcal{T}_2 \subseteq \mathcal{T}_1$ be a coarser topology on X. Then $(X, \mathcal{T}_2)$ is compact.

Problem 2.8

Use Theorem 2.2 to prove this corollary.

Hint Consider the identity function on X.

2.3 Products of compact topological spaces

One important construction in topology is the product topology. For example, it is how we get the Euclidean topology on $\mathbb{R}^n$ from that on $\mathbb{R}$. The product construction preserves a number of properties and here we show that it preserves compactness — *the product of two compact spaces is compact*.

The product topology was introduced in *Unit A3*. Given two topological spaces $(X, \mathcal{T}_X)$ and $(Y, \mathcal{T}_Y)$, we construct the product topology $\mathcal{T}_{X \times Y}$ on $X \times Y$ by using bases. Recall that a collection $\mathcal{B}$ of subsets of X is a base for a topology $\mathcal{T}$ on X if

Unit A3, Section 5.

(B1) $\mathcal{B} \subseteq \mathcal{T}$;

(B2) each set in $\mathcal{T}$ can be written as a union of sets in $\mathcal{B}$.

If $(X, \mathcal{T}_X)$ and $(Y, \mathcal{T}_Y)$ are topological spaces, then the product topology $\mathcal{T}_{X \times Y}$ on $X \times Y$ is the topology obtained from the base

$$\mathcal{B}_{X \times Y} = \{U \times V : U \in \mathcal{T}_X, V \in \mathcal{T}_Y\}.$$

The following result is an immediate consequence of this definition and is used in our proof that the product of two topological spaces is compact.

Lemma 2.5

Let $(X, \mathcal{T}_X)$ and $(Y, \mathcal{T}_Y)$ be topological spaces, let $\mathcal{T}_{X \times Y}$ be the product topology on $X \times Y$ and let $W \in \mathcal{T}_{X \times Y}$. Then, for any point $(x, y) \in W$, there are sets $U \in \mathcal{T}_X$ and $V \in \mathcal{T}_Y$ such that

$$(x, y) \in U \times V \subseteq W.$$

We also use the following result, which we proved in *Unit A3*.

Unit A3, Theorem 5.6.

Lemma 2.6

Let $(X, \mathcal{T}_X)$ and $(Y, \mathcal{T}_Y)$ be topological spaces and let $\mathcal{T}_{X \times Y}$ be the product topology on $X \times Y$. Then the projection functions $p_1 : X \times Y \to X$ and $p_2 : X \times Y \to Y$, given by $p_1(x, y) = x$ and $p_2(x, y) = y$, are $(\mathcal{T}_{X \times Y}, \mathcal{T}_X)$-continuous and $(\mathcal{T}_{X \times Y}, \mathcal{T}_Y)$-continuous, respectively.

We can now prove the main theorem on compactness for product spaces.

Theorem 2.7 Tikhonov's Theorem

Let $(X, \mathcal{T}_X)$ and $(Y, \mathcal{T}_Y)$ be non-empty topological spaces and let $\mathcal{T}_{X \times Y}$ be the product topology on $X \times Y$. Then $(X \times Y, \mathcal{T}_{X \times Y})$ is compact if and only if $(X, \mathcal{T}_X)$ and $(Y, \mathcal{T}_Y)$ are compact.

Andrei Nikolaevich Tikhonov (1906–93) was born in Gzhatska in Russia. Apart from this theorem, his main contribution to topology was discovering how to define a topology on an arbitrary product of topological spaces. There are many versions of his name — for example, Tychonoff.

Proof First suppose that $(X \times Y, \mathcal{T}_{X \times Y})$ is compact.
Now $X = p_1(X \times Y)$, and, by Lemma 2.6, p_1 is $(\mathcal{T}_{X \times Y}, \mathcal{T}_X)$-continuous. Thus X is the continuous image of a compact set and so, by Theorem 2.2, is compact. Similarly, $Y = p_2(X \times Y)$ is compact.

The proof that if $(X, \mathcal{T}_X)$ and $(Y, \mathcal{T}_Y)$ are compact then $(X \times Y, \mathcal{T}_{X \times Y})$ is compact is rather harder, and is similar in spirit to the proof that $[0,1]$ is compact.

Theorem 1.1.

Suppose that $(X, \mathcal{T}_X)$ and $(Y, \mathcal{T}_Y)$ are compact, and let $\mathcal{S}$ be an open cover of $X \times Y$.

We shall say that a subset $A \subseteq X$ is *good* (for $\mathcal{S}$) if there is a finite subcover from $\mathcal{S}$ of $A \times Y$. Our aim is to show that X is good and hence that $X \times Y$ is compact.

We break the proof into three steps.

1   For each $x \in X$, there is a good neighbourhood of x

This is the heart of the proof.

Let $x \in X$. Since $\mathcal{S}$ is a cover of $X \times Y$, we can find, for each $y \in Y$, an open set $W_y \in \mathcal{S}$ for which $(x, y) \in W_y$. Hence, by Lemma 2.5, we can find open sets $U_y \in \mathcal{T}_X$ and $V_y \in \mathcal{T}_Y$ such that

$$(x, y) \in U_y \times V_y \subseteq W_y.$$

The collection $\mathcal{R} = \{V_y : y \in Y\}$ is an open cover of the compact set Y, and so we can find a finite subcover $\{V_{y_1}, V_{y_2}, \ldots, V_{y_n}\}$ of Y from $\mathcal{R}$.

We now find a neighbourhood U of x such that $U \times Y$ is covered by $\{W_{y_1}, W_{y_2}, \ldots, W_{y_n}\}$. Set

$$U = U_{y_1} \cap U_{y_2} \cap \cdots \cap U_{y_n}.$$

U is open.

Then $x \in U$ and $U \subseteq U_{y_i}$, for each i. Now

$$U \times Y \subseteq U \times \bigcup_{i=1}^{n} V_{y_i} = \bigcup_{i=1}^{n} (U \times V_{y_i}).$$

But

$$U \times V_{y_i} \subseteq U_{y_i} \times V_{y_i} \subseteq W_{y_i}.$$

Therefore

$$U \times Y \subseteq \bigcup_{i=1}^{n} W_{y_i}.$$

Hence $U \times Y$ is covered by $\{W_{y_1}, W_{y_2}, \ldots, W_{y_n}\}$, and so U is a neighbourhood of x that is good.

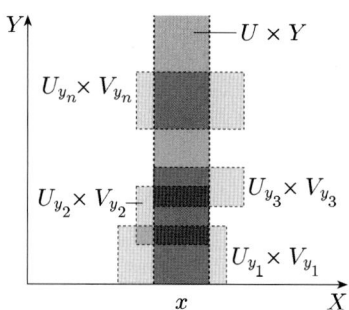

Figure 2.1

2   The union of a finite number of good subsets is good

Suppose that $A_1, A_2, \ldots, A_n \subseteq X$ are good. Then each set $A_i \times Y$ has a finite subcover $\mathcal{R}_i$ from $\mathcal{S}$. But then

$$(A_1 \cup A_2 \cup \cdots \cup A_n) \times Y = \bigcup_{i=1}^{n} (A_i \times Y)$$

is covered by the finite subcollection $\mathcal{R}_1 \cup \mathcal{R}_2 \cup \cdots \cup \mathcal{R}_n$ from $\mathcal{S}$, and so $A_1 \cup A_2 \cup \cdots \cup A_n$ is good.

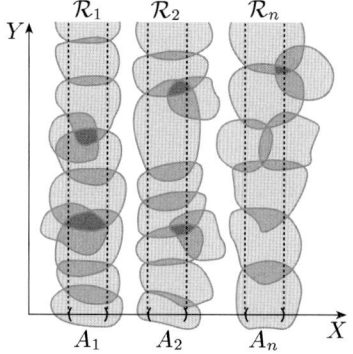

Figure 2.2

3   The set X is good

For each $x \in X$, Step 1 implies that we can find a neighbourhood U_x of x that is good. Thus

$$\mathcal{P} = \{U_x : x \in X \text{ and } U_x \text{ is good}\}$$

is an open cover of X. The compactness of X now implies that we can find a finite subcover $\{U_{x_1}, U_{x_2}, \ldots, U_{x_n}\}$ of X from $\mathcal{P}$. By Step 2, since each set U_{x_i} is good, so is their union, since it is a finite union of good sets. But $X = \bigcup_{i=1}^{n} U_{x_i}$, and so X is good (for $\mathcal{S}$).   ∎

Remarks

(i) We can extend this result to the product of any finite number of topological spaces: the product is compact if and only if each factor is compact. We use this later when we investigate compactness in $\mathbb{R}^n$.

This is an important result.

(ii) Even more remarkable is the fact that the result is true for the product of *arbitrarily* many spaces — even uncountably many. Although we have not discussed arbitrary products, we can at least see what a countably infinite product entails. Informally, if we have a topological space $(X_n, \mathcal{T}_n)$ for each $n \in \mathbb{N}$, then an element of the product space is an object of the form $(x_1, x_2, \ldots, x_n, \ldots)$, where $x_n \in X_n$. This is a generalization of a sequence in X: for such a sequence, every X_n is equal to X.

This was also proved by Tikhonov. When mathematicians refer to Tikhonov's Theorem, it is usually the unrestricted product form that they mean.

Problem 2.9

Show that $K = \{(x,y) : 0 \leq x \leq 1, 0 \leq y \leq 1\}$ is a compact subset of $\mathbb{R}^2$ for the Euclidean topology.

3 Compactness in Hausdorff spaces

> After working through this section, you should be able to:
> ▶ explain what is meant by a *Hausdorff space*;
> ▶ decide whether a given topological space is Hausdorff;
> ▶ decide whether a given subset of a Hausdorff space is compact.

In this section, we look at a special class of topological spaces known as *Hausdorff spaces*. For these spaces, compact sets have many useful properties in addition to those that they possess in a general topological space. Many of the spaces that we are interested in (including Euclidean spaces) are Hausdorff.

3.1 Hausdorff spaces

We begin by defining Hausdorff spaces. We first met this notion in *Unit B1*, but discussed it only briefly there.

Definition

A topological space $(X, \mathcal{T})$ is a **Hausdorff space** if each pair of distinct points in X has disjoint neighbourhoods — in other words, if $x, y \in X$ with $x \neq y$, then there are disjoint open sets U and V with $x \in U$ and $y \in V$.

Felix Hausdorff was born in Breslau, Germany (now Wrocław, Poland) in 1868. His book on set theory, published in 1914, contains probably the first statement of the axioms for a *topological space*. His work on dimension in topological spaces is now the core of the treatment of fractals. He took his own life in 1942 to avoid deportation to a concentration camp.

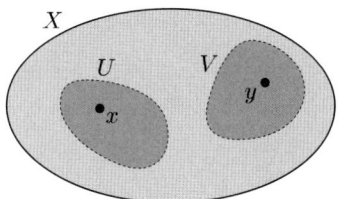

Figure 3.1

Remarks

(i) An effective mnemonic is to think of the two points x and y as being 'housed orff' from one another.
(ii) If X is empty, or contains only one point, then X is automatically Hausdorff.
(iii) This definition gives an example of a *separation axiom*. In Hausdorff spaces, distinct points can be 'separated' by disjoint neighbourhoods. This means that the subspace topology for any pair of points in X is the discrete topology.

Worked problem 3.1 *Discrete topology*

Let X be a set and let $\mathcal{T}$ be the discrete topology on X. Show that $(X, \mathcal{T})$ is Hausdorff.

Solution

If X is empty or contains only one point, $(X, \mathcal{T})$ is automatically Hausdorff. Otherwise, let $x, y \in X$ with $x \neq y$. Since every subset of X is open, $\{x\}$ and $\{y\}$ are disjoint open sets with $x \in \{x\}$ and $y \in \{y\}$. Thus $(X, \mathcal{T})$ is Hausdorff. ∎

Problem 3.1 Indiscrete topology

Let X be a set containing at least two points and let $\mathcal{T}$ be the indiscrete topology on X. Show that $(X, \mathcal{T})$ is not Hausdorff.

We now show that there are many Hausdorff spaces.

Theorem 3.1

Each metric space is Hausdorff.

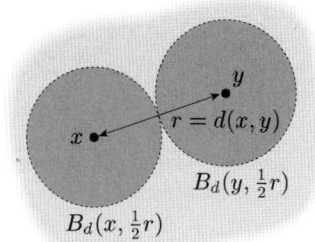

Figure 3.2

Proof Let (X, d) be a metric space. If X is empty or contains just one point, there is nothing to prove. Otherwise, consider $x, y \in X$ with $x \neq y$. Then $d(x, y) > 0$. So, if $r = d(x, y)$, then $B_d(x, \tfrac{1}{2}r)$ and $B_d(y, \tfrac{1}{2}r)$ are disjoint neighbourhoods of x and y, respectively. ∎

In particular, $(\mathbb{R}^n, \mathcal{T}(d^{(n)}))$ for $n \in \mathbb{N}$, $(C[0,1], \mathcal{T}(d_{\max}))$ and the Cantor space $(\mathbf{C}, \mathcal{T}(d_\mathbf{C}))$ are all Hausdorff spaces.

See *Units A2* and *A3* for the definitions of these spaces.

We have seen that for $\mathbb{R}$, with its Euclidean topology, all one-point sets are closed but not open. For the indiscrete topology, in contrast, if X has at least two points, then $\{x\}$ is neither open nor closed for each $x \in X$. We now prove the following general result.

See *Units A2* and *A4*.

See *Units A3* and *A4*.

Theorem 3.2

Let $(X, \mathcal{T})$ be a Hausdorff space. For each $x \in X$, $\{x\}$ is a closed set.

Proof Let $(X, \mathcal{T})$ be a Hausdorff space. If X is empty or contains just one point, there is nothing to prove. Otherwise, suppose that X contains at least two points, and let $x \in X$ and let $A = X - \{x\} \neq \varnothing$. We show that A contains a neighbourhood of each of its points, from which it follows that A is open, and so $\{x\}$ (being the complement of an open set) is closed.

Unit A4, Theorem 2.1.

Suppose that $y \in A$. Since $y \neq x$, the Hausdorff property of X implies that there are disjoint neighbourhoods U of x and V of y. Since U and V are disjoint, $x \notin V$ and so $V \subseteq A$. Hence, since y is an arbitrary point of A, the set A contains a neighbourhood of each of its points, and so is open. ∎

Problem 3.2

Let X be a set containing at least two points, let $a \in X$ and let $\mathcal{T}_a$ be the a-deleted-point topology on X. Show that $(X, \mathcal{T}_a)$ is not a Hausdorff space.

U is open if either $U = X$ or $a \notin U$.

Hint Show that $\{x\}$ is not a closed set for each $x \in X$ such that $x \neq a$.

A natural question to ask is whether subspaces of a Hausdorff space are Hausdorff.

Theorem 3.3

Let $(A, \mathcal{T}_A)$ be a subspace of a Hausdorff space $(X, \mathcal{T})$. Then $(A, \mathcal{T}_A)$ is Hausdorff.

$\mathcal{T}_A$ denotes the subspace topology on A inherited from $\mathcal{T}$.

Problem 3.3

Prove Theorem 3.3.

Remark

It follows from Theorem 3.3 that any subset of $(\mathbb{R}^n, \mathcal{T}(d^{(n)}))$ with the subspace topology is a Hausdorff space.

We now examine how Hausdorff spaces and continuous functions interact.

Theorem 3.4

Let $(X, \mathcal{T}_X)$ and $(Y, \mathcal{T}_Y)$ be topological spaces, and let $f: X \to Y$ be $(\mathcal{T}_X, \mathcal{T}_Y)$-continuous and one–one. If $(Y, \mathcal{T}_Y)$ is Hausdorff, then $(X, \mathcal{T}_X)$ is Hausdorff.

This result is not necessarily true if f is not one–one.

Proof If X is empty or contains just one point, there is nothing to prove. Otherwise, let $x, y \in X$ with $x \neq y$. We must show that there are disjoint neighbourhoods of x and y.

Since f is one–one, $f(x) \neq f(y)$. Hence, since Y is Hausdorff, there are disjoint neighbourhoods U of $f(x)$ and V of $f(y)$. Since f is continuous, $f^{-1}(U)$ and $f^{-1}(V)$ are neighbourhoods of x and y, respectively. We need show only that they are disjoint.

Suppose that they are not disjoint, and that $z \in f^{-1}(U) \cap f^{-1}(V)$. Then $f(z) \in U$ and $f(z) \in V$, contradicting the fact that U and V are disjoint. Hence no such point z exists.

Thus $(X, \mathcal{T}_X)$ is Hausdorff. ∎

Since a homeomorphism and its inverse function are both one–one and continuous, it follows that

if $(X, \mathcal{T}_X)$ and $(Y, \mathcal{T}_Y)$ are homeomorphic topological spaces, and one of them is Hausdorff, then so is the other.

We can restate this result as follows.

Corollary 3.5

Hausdorffness is a topological invariant.

We now show that the product of two Hausdorff spaces is Hausdorff. The proof is more straightforward than the corresponding result for compact spaces.

Theorem 3.6

The product of two Hausdorff spaces is Hausdorff.

Proof Let $(X, \mathcal{T}_X)$ and $(Y, \mathcal{T}_Y)$ be Hausdorff spaces, and let $X \times Y$ carry its product topology $\mathcal{T}_{X \times Y}$. If X and Y both contain at most one point, then so does $X \times Y$ and hence there is nothing to prove. Otherwise, let (x_1, y_1) and (x_2, y_2) be distinct points in $X \times Y$. Then either $x_1 \neq x_2$ or $y_1 \neq y_2$ (or both).

Suppose that $x_1 \neq x_2$. Since X is a Hausdorff space, there are disjoint $\mathcal{T}_X$-neighbourhoods U_1 of x_1 and U_2 of x_2. Then $U_1 \times Y$ is a neighbourhood of (x_1, y_1) and $U_2 \times Y$ is a neighbourhood of (x_2, y_2). Moreover,

$$(U_1 \times Y) \cap (U_2 \times Y) = (U_1 \cap U_2) \times Y = \varnothing \times Y = \varnothing,$$

and so these are disjoint neighbourhoods.

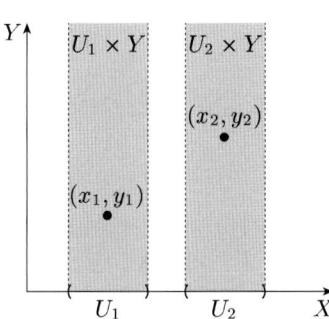

Figure 3.3 $x_1 \neq x_2$

Now suppose that $y_1 \neq y_2$. Since Y is Hausdorff, there are disjoint $\mathcal{T}_Y$-neighbourhoods V_1 of y_1 and V_2 of y_2. Then $X \times V_1$ is a neighbourhood of (x_1, y_1) and $X \times V_2$ is a neighbourhood of (x_2, y_2). Moreover,

$$(X \times V_1) \cap (X \times V_2) = X \times (V_1 \cap V_2) = X \times \varnothing = \varnothing,$$

and so these are disjoint neighbourhoods.

Thus $(X \times Y, \mathcal{T}_{X \times Y})$ is a Hausdorff space. ∎

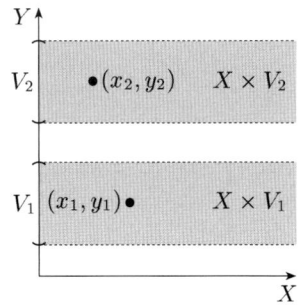

Figure 3.4 $y_1 \neq y_2$

This result generalizes in the usual way: the product of a finite number of Hausdorff spaces is also Hausdorff.

3.2 Compact sets in Hausdorff spaces

When introducing Hausdorff spaces, we stated that compact subsets of Hausdorff spaces have further useful properties that are not true for compact subsets of a general topological space. We now describe some of these properties.

We saw in Theorem 2.1 that a closed subset of a compact space is a compact set. In the next theorem, we state a partial converse of this result — a compact subset of a Hausdorff space is always a closed set.

Theorem 3.7

Let $(X, \mathcal{T})$ be a Hausdorff space and let $K \subseteq X$ be compact. Then K is closed.

Proof It is enough to show that K^c is open. In order to do this, we show that each point in K^c has a neighbourhood contained in K^c, and hence K^c is open, by Theorem 2.1 of *Unit A4*.

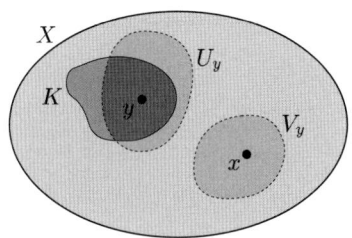

Figure 3.5

If $K^c = \varnothing$, then it is open. So suppose $K^c \neq \varnothing$ and fix $x \in K^c$, and let $y \in K$. Since X is Hausdorff, there are disjoint neighbourhoods U_y of y and V_y of x. Thus

$$\mathcal{S} = \{U_y : y \in K\}$$

is an open cover of K, and so has a finite subcover $\{U_{y_1}, U_{y_2}, \ldots, U_{y_n}\}$ of K. Thus

$$K \subseteq \bigcup_{i=1}^{n} U_{y_i} = U, \text{ say.}$$

Let

$$V = \bigcap_{i=1}^{n} V_{y_i}.$$

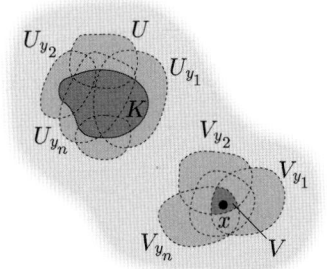

Figure 3.6

Then $x \in V$, and V is open since it is the intersection of a finite number of open sets. If we can show that V is disjoint from K, then V is a neighbourhood of x that is contained in K^c.

Using the Distributive Laws for sets, we have

Unit A3, Theorem 2.6.

$$U \cap V = \left(\bigcup_{i=1}^n U_{y_i}\right) \cap V = \bigcup_{i=1}^n (U_{y_i} \cap V) = \bigcup_{i=1}^n \left(U_{y_i} \cap \left(\bigcap_{j=1}^n V_{y_j}\right)\right)$$
$$\subseteq \bigcup_{i=1}^n (U_{y_i} \cap V_{y_i}) = \varnothing.$$

Hence U and V are disjoint, and so $V \subseteq U^c \subseteq K^c$, as required. ∎

We know from Theorem 3.1 that all metric spaces are Hausdorff. It follows from Theorem 3.7 that compact subsets of metric spaces are closed. We now show that such a set must also be *bounded*.

Definition

Let (X, d) be a metric space and let $A \subseteq X$. Then A is **bounded** if there is $M > 0$ such that $d(x, y) \leq M$ for all $x, y \in A$.

For the Euclidean topology on $\mathbb{R}$, this definition is equivalent to the definition that we gave in *Unit A1*.

In *Unit A1* we said that $A \subset \mathbb{R}$ is bounded if there is $M > 0$ such that $|x| \leq M$ for all $x \in A$.

Problem 3.4

Show that the definition of a bounded subset of $\mathbb{R}$ given in *Unit A1* is equivalent to the definition above for the Euclidean metric on $\mathbb{R}$.

Theorem 3.8

Let (X, d) be a metric space and let $K \subseteq X$ be compact. Then K is closed and bounded.

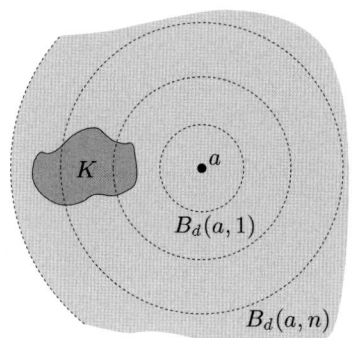

Proof If $K = \varnothing$, then K is closed and bounded, so suppose $K \neq \varnothing$. We have already observed that metric spaces are Hausdorff (by Theorem 3.1). It follows from Theorem 3.7 that K is closed.

Figure 3.7

It remains to show that K is bounded. Suppose that $a \in X$, and consider the open cover of X given by $\mathcal{S} = \{B_d(a, n) : n \in \mathbb{N}\}$. This is also an open cover of K, since $K \subseteq X$. Since K is compact, there is a finite subcover of K from $\mathcal{S}$. Now the sets making up $\mathcal{S}$ are nested, so this finite subcover reduces to a single-set subcover $\{B_d(a, N)\}$, for some $N \in \mathbb{N}$. So $K \subseteq B_d(a, N)$.

Thus, using the Triangle Inequality, we have

$$d(x, y) \leq d(x, a) + d(a, y) < 2N,$$

for all $x, y \in K$. This proves that K is bounded. ∎

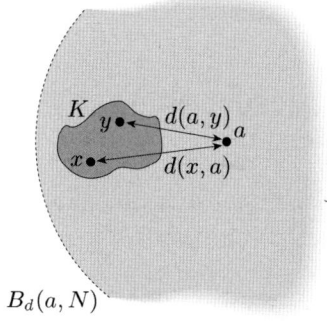

Figure 3.8

In Section 4, we shall see that in Euclidean spaces the converse to this result is true, and so the compact sets in Euclidean spaces are precisely those sets that are both closed and bounded. However, the converse to Theorem 3.8 is not true in general — a subset of a metric space can be closed and bounded without being compact.

Problem 3.5

For the discrete topology on $\mathbb{R}$, show that $[0,1]$ is closed and bounded but not compact.

Another useful consequence of Theorem 3.7 concerns the intersection of compact sets.

Theorem 3.9

Let $(X, \mathcal{T})$ be a Hausdorff space. Then the intersection of any collection of compact subsets of X is compact.

Proof Let $\mathcal{F}$ be a collection of compact subsets of X. We must show that $\bigcap_{K \in \mathcal{F}} K$ is compact. Theorem 3.7 implies that K is closed, for each $K \in \mathcal{F}$, and hence $\bigcap_{K \in \mathcal{F}} K$ is also a closed set (by Theorem 1.4 of *Unit A4*). It follows that $\bigcap_{K \in \mathcal{F}} K$ is a closed subset (for the subspace topology) of any one of the compact sets K in the collection $\mathcal{F}$, and so, by Theorem 2.1, $\bigcap_{K \in \mathcal{F}} K$ is compact. ∎

You may be surprised to learn that there are topological spaces in which the intersection of two compact sets need not itself be compact. This result implies that such spaces cannot be Hausdorff. We now give an example of such a space.

Example 3.1

Let $(X, \mathcal{T})$ be a topological space, where $X = \mathbb{R} \times \{0, 1\}$ and $\mathcal{T}$ is the product topology on X formed from the Euclidean topology on $\mathbb{R}$ and the indiscrete topology on $\{0, 1\}$. It can be shown that the sets in $\mathcal{T}$ are of the form $U \times \{0, 1\}$, where U is an open subset of $\mathbb{R}$ for the Euclidean topology on $\mathbb{R}$.

Now let

$$A = ([0,1] \times \{0\}) \cup ((0,1) \times \{1\}),$$
$$B = ((0,1) \times \{0\}) \cup ([0,1] \times \{1\}),$$

as illustrated in Figure 3.9. We show that A and B are compact subsets of X, but that their intersection is not.

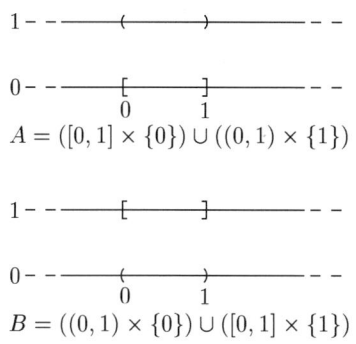

Figure 3.9

To see why A is compact, let $\mathcal{S} = \{U_i \times \{0, 1\}\}$ be an open cover of A. Then $\{U_i\}$ is an open cover of $[0, 1]$. Since $[0, 1]$ is compact, there is a finite subcover $\{U_1, U_2, \ldots, U_n\}$ of $[0, 1]$. It follows that $\{U_i \times \{0, 1\} : 1 \leq i \leq n\}$ is a finite subcover of A from $\mathcal{S}$ and so A is compact. Similar arguments show that B is compact.

The fact that $A \cap B = ((0,1) \times \{0\}) \cup ((0,1) \times \{1\})$ is not compact follows from the fact that $(0, 1)$ is not compact. We omit the details. ∎

In *Unit C5*, we investigate some basic properties of fractal sets. You will see that many such sets can be defined as intersections of compact subsets of Hausdorff spaces. Theorem 3.9 guarantees that the resulting intersection is compact. However, the intersection can be empty, particularly if a collection $\mathcal{F}$ of compact sets is large. For example, if $\mathcal{F} = \{\{x\} : x \in [0,1]\}$, then $\bigcap_{K \in \mathcal{F}} K = \emptyset$. Fortunately, there is one situation, useful in the study of fractals, where the intersection of a collection of compact sets must be non-empty in a Hausdorff space.

Theorem 3.10

Let $(X, \mathcal{T})$ be a Hausdorff space, and let K_n ($n = 1, 2, 3, \ldots$) be non-empty compact sets in X for which
$$K_1 \supseteq K_2 \supseteq K_3 \supseteq \cdots \supseteq K_n \supseteq K_{n+1} \supseteq \cdots.$$
Then $\bigcap_{n=1}^{\infty} K_n$ is a non-empty compact set.

Proof Theorem 3.9 implies that $\bigcap_{n=1}^{\infty} K_n$ is compact. It remains to show that the intersection is non-empty.

Since X is Hausdorff, it follows from Theorem 3.7 that each K_n is closed. Thus K_n^c is open, for each n. Moreover, by De Morgan's Second Law,

Unit A3, Theorem 2.10.

$$\bigcup_{n=1}^{\infty} K_n^c = \left(\bigcap_{n=1}^{\infty} K_n \right)^c.$$

We assume that $\bigcap_{n=1}^{\infty} K_n = \varnothing$ and use proof by contradiction. Then $\{K_n^c : n \in \mathbb{N}\}$ is an open cover of X, and so (in particular) is an open cover of K_1. Since K_1 is compact, there is a number $N \in \mathbb{N}$ for which

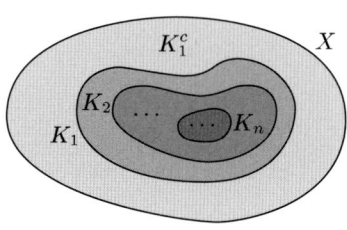

$$K_1 \subseteq \bigcup_{n=1}^{N} K_n^c.$$

Figure 3.10

Using De Morgan's Second Law again, we see that
$$\bigcup_{n=1}^{N} K_n^c = \left(\bigcap_{n=1}^{N} K_n \right)^c = K_N^c.$$

Therefore $K_1 \subseteq K_N^c$. Hence
$$K_N \subseteq K_1 \subseteq K_N^c.$$

We deduce that K_N is empty — a contradiction. Hence $\bigcap_{n=1}^{\infty} K_n \neq \varnothing$, as required. ∎

Example 3.2

Recall that the Cantor space **C** is the space of all infinite sequences of 0s and 1s:

Unit A2, Subsection 2.2.

$$\mathbf{C} = \{(a_n) : a_n \in \{0, 1\} \text{ for all } n \in \mathbb{N}\}.$$

We defined the *Cantor distance* $d_{\mathbf{C}} : \mathbf{C} \times \mathbf{C} \to \mathbb{R}$ by
$$d_{\mathbf{C}}(\mathbf{x}, \mathbf{y}) = \begin{cases} 0 & \text{if } \mathbf{x} = \mathbf{y}, \\ 2^{-n} & \text{if } \mathbf{x} \text{ and } \mathbf{y} \text{ first differ at the } n\text{th term.} \end{cases}$$

There is a nice way of visualizing the Cantor space that makes use of Theorem 3.10.

We work in $\mathbb{R}$ with the Euclidean topology. Let $I_0 = [0, 1]$ be the unit interval: this is a compact set, as we saw in Theorem 1.1. Let I_1 consist of the two intervals that you obtain from I_0 if you omit the middle-third of I_0, the open interval $(\frac{1}{3}, \frac{2}{3})$. Thus
$$I_1 = [0, \tfrac{1}{3}] \cup [\tfrac{2}{3}, 1].$$

This is a closed subset of I_0 and so, by Theorem 2.1, is a compact set. Let I_2 be the four intervals that you get if you omit the (open) middle-third of each interval making up I_1. So
$$I_2 = [0, \tfrac{1}{9}] \cup [\tfrac{2}{9}, \tfrac{1}{3}] \cup [\tfrac{2}{3}, \tfrac{7}{9}] \cup [\tfrac{8}{9}, 1].$$

Again, since I_2 is a closed subset of the compact set I_0, it is compact. Proceeding similarly, for $n > 1$, define I_n to be the set that you obtain by omitting the (open) middle-thirds of the intervals that make up I_{n-1}. Then I_n is a compact set consisting of 2^n intervals of width 3^{-n} (see Figure 3.11).

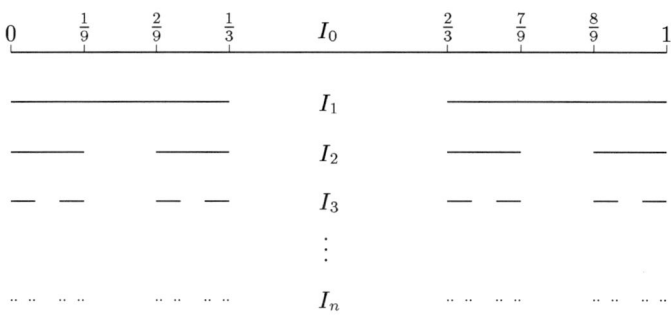

Figure 3.11

Moreover,
$$I_0 = [0,1] \supseteq I_1 \supseteq I_2 \supseteq \cdots \supseteq I_n \supseteq I_{n+1} \supseteq \cdots.$$

Theorem 3.10 now implies that
$$C = \bigcap_{n=1}^{\infty} I_n$$
is a non-empty compact set. This set is known as the *middle-third Cantor set*.

We explore this topic further in *Unit C5*.

The connection with the Cantor space **C** can be seen as follows: an element $\mathbf{x} \in \mathbf{C}$ can be thought of as giving the (unique) address of a point in C, as illustrated in Figure 3.12.

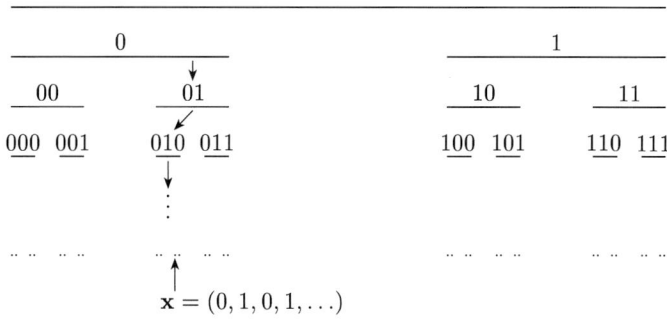

Figure 3.12 Mapping **C** to C.

The resulting map is $(\mathcal{T}(d_{\mathbf{C}}), \mathcal{T}(d^{(2)}))$-continuous. ∎

4 Compactness in Euclidean spaces

After working through this section, you should be able to:
▶ give and use an alternative characterization of compact sets in Euclidean spaces;
▶ understand and use a generalization of the Extreme Value Theorem.

In the last section we saw that each compact subset of a metric space is closed and bounded. We also observed that the converse statement is not true in general. In this section we show that, in Euclidean spaces,

a set is compact if and only if it is both closed and bounded.

Thus, in Euclidean spaces, the implications go in both directions.

This classification of the compact subsets of Euclidean spaces enables us to achieve one of the main objectives of this unit and prove a generalization of the Extreme Value Theorem.

4.1 Compact subsets of Euclidean spaces

We have already done most of the work to prove that the compact subsets of $\mathbb{R}$ with its Euclidean topology coincide with the closed and bounded subsets, so we can proceed directly to the following result.

Recall that a set $A \subset \mathbb{R}$ is *bounded* if there is $M > 0$ such that $|x| \leq M$ for all $x \in A$.

Theorem 4.1 Heine–Borel–Lebesgue Theorem for $\mathbb{R}$

A subset of $\mathbb{R}$ is $\mathcal{T}(d^{(1)})$-compact if and only if it is closed and bounded.

Proof Suppose that $K \subseteq \mathbb{R}$ is compact. It follows from Theorem 3.8 that K is closed and bounded.

Now suppose that $K \subseteq \mathbb{R}$ is closed and bounded. Since K is bounded, we can find $M > 0$ such that $K \subseteq [-M, M]$. Now $[-M, M]$ is compact. Hence K is a closed subset of a compact set and so, by Theorem 2.1, K is compact. ■

In Example 2.3 we showed that any closed and bounded interval is compact.

Problem 4.1

Which of the following subsets of $\mathbb{R}$ are $\mathcal{T}(d^{(1)})$-compact?

(a) $\{0\} \cup \{\frac{1}{n} : n \in \mathbb{N}\}$ (b) $[0, 1] \cap \mathbb{Q}$ (c) $[0, 1] \cup [e, \pi] \cup \{100\}$
(d) $[0, \infty)$

We now look at the generalization of Theorem 4.1 to higher-dimensional Euclidean spaces, such as the plane.

Theorem 4.2 Heine–Borel–Lebesgue Theorem for $\mathbb{R}^n$

A subset of $\mathbb{R}^n$ is $\mathcal{T}(d^{(n)})$-compact if and only if it is closed and bounded.

Proof The first part of the proof is similar to that for Theorem 4.1.

Suppose that $K \subseteq \mathbb{R}^n$ is compact. It follows from Theorem 3.8 that K is closed and bounded.

Now suppose that $K \subseteq \mathbb{R}^n$ is closed and bounded. Since K is bounded, $p_j(K)$ is bounded for each $j = 1, 2, \ldots, n$, where p_j is the usual projection function onto the jth coordinate axis. So we can find $M_j > 0$ such that

$$p_j(K) \subseteq [-M_j, M_j], \quad j = 1, 2, \ldots, n.$$

Hence

$$K \subseteq [-M_1, M_1] \times [-M_2, M_2] \times \cdots \times [-M_n, M_n].$$

Each of the intervals $[-M_j, M_j]$ is a compact subset of $\mathbb{R}$. Thus, by Tikhonov's Theorem, the product of the intervals is also compact. Hence K is a closed subset of a compact set, and so is compact, by Theorem 2.1. ∎

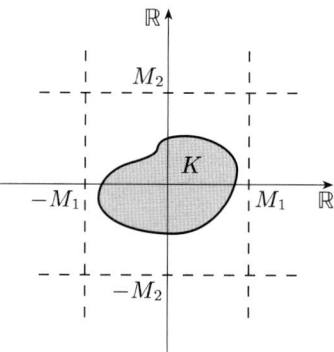

Figure 4.1

Remark

The set $\mathbb{R}^n$ is not bounded for the Euclidean metric $d^{(n)}$ and so $(\mathbb{R}^n, \mathcal{T}(d^{(n)}))$ is not compact.

Problem 4.2

Which of the following subsets of $\mathbb{R}^2$ are $\mathcal{T}(d^{(2)})$-compact?

(a) $[0,1] \times [1,4]$ (b) $\{(0,0), (1,0), (0,1), (1,1)\}$

(c) $\{(x,y) : x^2 + y^2 < 1\}$ (d) $\{(x,y) : x^2 + y^2 = 1\}$

4.2 The Extreme Value Theorem

We now use our classification of the compact subsets of the real line to prove a general version of the Extreme Value Theorem.

> **Theorem 4.3 General Extreme Value Theorem**
>
> Let $(X, \mathcal{T})$ be a non-empty compact topological space, and let $f \colon X \to \mathbb{R}$ be $(\mathcal{T}, \mathcal{T}(d^{(1)}))$-continuous. Then there are $c, d \in X$ such that
>
> $$f(c) \leq f(x) \leq f(d) \quad \text{for all } x \in X.$$

Proof Since X is compact and f is continuous, Theorem 2.2 implies that $f(X)$ is a compact subset of $\mathbb{R}$. Hence, by Theorem 4.1, $f(X)$ is a closed and bounded subset of $\mathbb{R}$.

Since $f(X)$ is bounded, it has a least upper bound M and a greatest lower bound m. From the definition of a least upper bound, we can deduce that any neighbourhood of M must intersect $f(X)$. Similarly any neighbourhood of m must intersect $f(X)$. Thus M and m are closure points of $f(X)$. But $f(X)$ is closed. We deduce from Theorem 2.4 of *Unit A4* that M and m are both in $f(X)$. Thus there are $c \in X$ and $d \in X$ with $f(c) = m$ and $f(d) = M$. It follows from the definition of m and M that

$$f(c) \leq f(x) \leq f(d) \quad \text{for all } x \in X.$$ ∎

It follows from Theorem 4.3 that any continuous function from a compact subset of $\mathbb{R}^n$ to $\mathbb{R}$ has a least upper bound and a greatest lower bound, and attains them both: in other words, it has a maximum value and a minimum value.

If the function is continuous but the domain is not compact, then this need not be the case. For example, consider $f\colon \mathbb{R} \to \mathbb{R}$ given by $f(x) = x$. This function is certainly continuous, but on the open unit interval $(0,1)$ it attains neither its least upper bound at $f(1) = 1$ nor its greatest lower bound at $f(0) = 0$: in other words, it has neither a maximum value nor a minimum value.

Worked problem 4.1

Let $f\colon \mathbb{R}^2 \to \mathbb{R}$ be given by
$$f(x,y) = \sqrt{x^2 + y^2}.$$
Show that f attains a maximum value on $B_{d^{(2)}}[(0,0),1]$.

Solution

In order to use Theorem 4.3 to deduce the existence of a maximum value, we must verify that f is continuous on its domain and that $B_{d^{(2)}}[(0,0),1]$ is a compact set.

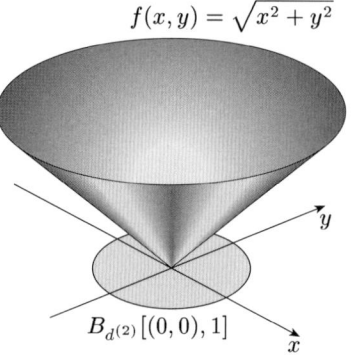

Figure 4.2

We saw in Worked problem 4.4 of *Unit A1* that f is continuous on the whole of $\mathbb{R}^2$, and so, by the Restriction Rule, is continuous on $B_{d^{(2)}}[(0,0),1]$.

The ball $B_{d^{(2)}}[(0,0),1]$ is closed and bounded. Hence, by Theorem 4.2, it is compact.

Thus both conditions of the General Extreme Value Theorem are satisfied, and we conclude that the function f attains a maximum value on $B_{d^{(2)}}[(0,0),1]$. ∎

Problem 4.3

Let $f\colon \mathbb{R}^2 \to \mathbb{R}$ be given by
$$f(x,y) = y\sin(\pi x) + x\cos(\pi y).$$
Show that f attains a minimum value on $\{(x,y) : 0 \le x, y \le 1\}$.

The General Extreme Value Theorem is a powerful result since it guarantees the existence of a maximum value and a minimum value of a continuous function on a compact domain without our having to know where these points are.

5 Surfaces and compactness

After working through this section, you should be able to:
▶ appreciate that the surfaces discussed in Block B of the course are compact Hausdorff spaces.

We conclude this unit by discussing the relationship between the surfaces that we studied in Block B and the definition of compactness that we introduced in Section 2.

This section is not assessed.

Recall that in *Unit B1*, our main concern was with *compact surfaces*, which we defined as follows.

Definition

A **compact surface** is a surface that can be obtained from a polygon by identifying edges.

We should check that this use of the word 'compact' for surfaces is consistent with its use in this unit. In other words, we must verify the following theorem.

Theorem 5.1

A compact surface is a compact topological space.

Proof Let P be a closed polygon in the plane, and let $\mathcal{T}$ denote the subspace topology on P inherited from the Euclidean topology on $\mathbb{R}$. Since P is a closed and bounded subset of the plane, it is compact, by Theorem 4.2.

Let $f: P \to P$ be a map that identifies (some) edges of P. Proceeding as in Subsection 4.1 of *Unit B1*, we define the *identification space* $I(P)$ of P under f to be the set of all identification classes $[x]$, where

$$[x] = \{y \in P : f(y) = f(x)\}.$$

We define a topology $\mathcal{T}_f$ on $I(P)$ by setting

$$\mathcal{T}_f = \{U \subseteq I(P) : p^{-1}(U) \in \mathcal{T}\},$$

where $p: P \to I(P)$ is given by $p(x) = [x]$. Moreover, the map p is $(\mathcal{T}, \mathcal{T}_f)$-continuous. The space $(I(P), \mathcal{T}_f)$ is (homeomorphic to) our compact surface. Since $I(P) = p(P)$, it follows from Theorem 2.2 that $I(P)$ is $\mathcal{T}_f$-compact as required. ∎

This follows from Theorem 4.1 of Unit B1.

In *Unit B1*, we also used the result that compact surfaces are Hausdorff. The proof of this depends on the form that disc-like neighbourhoods take after the identifications have been made — we omit the details.

Solutions to problems

1.1 One such subcollection is
$$\{(\tfrac{1}{10}(i-1), \tfrac{1}{10}(i+1)) : i = 0, 1, \ldots, 10\}.$$

1.2 (a) If $0 < x \le 1$, then there is a rational number $q > x$ and so $x \in (0, q)$. Also, $0 \in (-\tfrac{1}{100}, \tfrac{1}{100})$. Hence
$$[0,1] \subseteq \bigcup_{U \in \mathcal{S}_2} U.$$
Each interval $(0, q)$ is open, and so is the interval $(-\tfrac{1}{100}, \tfrac{1}{100})$. Thus $\mathcal{S}_2$ is an open cover of $[0, 1]$.

(b) One such subcollection is
$$\{(-\tfrac{1}{100}, \tfrac{1}{100}), (0, q)\},$$
where q is any fixed rational number *strictly* larger than 1.

1.3 One such collection is
$$\{\{0\}, (\tfrac{1}{2}, 1], (\tfrac{1}{3}, \tfrac{1}{2}], (\tfrac{1}{4}, \tfrac{1}{3}], \ldots, (\tfrac{1}{n+1}, \tfrac{1}{n}], \ldots\}.$$
If we omit even one set, the remaining collection is not a cover of $[0, 1]$. Indeed, whatever set we omit uncovers all the points of that set.

Another possibility is given by
$$\{\{x\} : x \in [0, 1]\}.$$
This is an uncountable cover of $[0, 1]$. If we omit just a single set $\{a\}$ in this cover then the point $a \in [0, 1]$ is no longer covered.

1.4 There are many open covers with this property. One such is
$$\mathcal{S} = \{(-1, 1), (0, 2), \ldots, (n-1, n+1), \ldots\}.$$
If $x \in [0, \infty)$, then x belongs to an interval of the form $(n-1, n+1)$, for some $n \in \mathbb{N}$. So $\mathcal{S}$ is a cover of $[0, \infty)$. Also, the intervals in $\mathcal{S}$ are all open. So $\mathcal{S}$ is an open cover of $[0, \infty)$. But, if
$$\{(n_1 - 1, n_1 + 1), \ldots, (n_k - 1, n_k + 1)\}$$
is a finite subcollection of $\mathcal{S}$, then it does not cover any point x for which
$$x \ge \max\{n_1, n_2, \ldots, n_k\} + 1.$$

2.1 Let $\mathcal{S}$ be an open cover of $\varnothing$. We must find a finite subcover of $\varnothing$ by sets from $\mathcal{S}$. The empty cover $\{\ \}$ is a finite subcover of $\mathcal{S}$ that covers $\varnothing$. We conclude that $\varnothing$ is compact.

2.2 Let $(X, \mathcal{T})$ be a topological space, and let A be a finite subset of X. If $A = \varnothing$, then A is compact by the result of Problem 2.1. So suppose instead that $A = \{a_1, a_2, \ldots, a_n\} \ne \varnothing$. If $\mathcal{S}$ is an open cover of A, then there is an open set $U_i \in \mathcal{S}$ such that $a_i \in U_i$, for each $i = 1, 2, \ldots, n$. Hence $\mathcal{R} = \{U_i : i = 1, 2, \ldots, n\}$ is a finite subcover of A from $\mathcal{S}$.

2.3 The only open sets that contain 0 are $(-1, 1)$, $[-1, 1)$, $(-1, 1]$ and $[-1, 1]$. Thus if $\mathcal{S}$ is an open cover of $[-1, 1]$, then it must contain at least one set containing the interval $(-1, 1)$, together with a set containing -1 and a set containing 1. By selecting these sets, we find a finite subcover of $[-1, 1]$, and so $[-1, 1]$ is compact.

An example of an open cover of $[-1, 0)$ which has no finite subcover is
$$\mathcal{S} = \{[-1, -\tfrac{1}{n}) : n \in \mathbb{N}\}.$$
If $\mathcal{R}$ is a finite subcollection of this cover, then there is an $N \in \mathbb{N}$ for which
$$\bigcup_{U \in \mathcal{R}} U \subseteq [-1, -\tfrac{1}{N})$$
and so the point $-\tfrac{1}{2N} \in [-1, 0)$ is not covered by $\mathcal{R}$. Hence $\mathcal{R}$ is not a finite subcover of $[-1, 0)$.

Thus $[-1, 0)$ is not compact.

2.4 An example of such a cover of $\mathbb{R}$ is
$$\mathcal{S} = \{(i-1, i+1) : i \in \mathbb{Z}\}.$$
To see that this has no finite subcover, it is sufficient to observe that any finite collection of intervals of the form $(i-1, i+1)$ is contained in $(-N, N)$ for some sufficiently large value of N, so cannot cover the whole real line.

Alternatively, we observe that if we omit just a single set $(i-1, i+1)$ from this collection, then the point i is no longer covered. Hence $\mathcal{S}$ has no proper subcover of $\mathbb{R}$, and so, in particular, contains no finite subcover of $\mathbb{R}$.

2.5 An example of such a cover of $\mathbb{R}^2$ is
$$\mathcal{S} = \{B((i,j), 1) : i, j \in \mathbb{Z}\}.$$
To see that this has no finite subcover, it is sufficient to observe that any finite collection of balls is contained within $B((0, 0), N)$ for some sufficiently large value of N, and so cannot cover the whole plane.

Alternatively, we observe that if we omit just a single set $B((i, j), 1)$ from this collection, then the point (i, j) is no longer covered. Hence $\mathcal{S}$ has no proper subcover of $\mathbb{R}^2$, and so, in particular, contains no finite subcover of $\mathbb{R}^2$.

2.6 In *Unit A4*, Worked problem 2.2, we showed that A is closed. So A is compact, by Theorem 2.1.

2.7 There are many possible examples here. For example, we know that $[0, 1]$ is compact whereas $(0, 1)$ is not compact, and the function $f : [0, 1] \to \mathbb{R}$ given by
$$f(x) = \begin{cases} \tfrac{1}{2} & \text{if } x = 0, \\ x & \text{if } x \in (0, 1), \\ \tfrac{1}{2} & \text{if } x = 1, \end{cases}$$
is discontinuous for the Euclidean topologies, with $f([0, 1]) = (0, 1)$.

2.8 Let id: $X \to X$ be the identity function id$(x) = x$. By Theorem 4.2 of *Unit A3*, id is $(\mathcal{T}_1, \mathcal{T}_2)$-continuous. Since X is $\mathcal{T}_1$-compact, it follows from Theorem 2.2 that id$(X) = X$ is $\mathcal{T}_2$-compact — that is, $(X, \mathcal{T}_2)$ is a compact topological space.

2.9 We can write $K = [0, 1] \times [0, 1]$. We know that $[0, 1]$ is compact and so K is the product of two compact spaces. Hence, by Theorem 2.7, K is compact.

3.1 Since X has at least two points, we can find $x, y \in X$ with $x \neq y$. For the indiscrete topology, the only possible neighbourhood of x is X, which is also a neighbourhood of y. Hence it is not possible to find disjoint neighbourhoods of x and y. Thus $(X, \mathcal{T})$ is not Hausdorff.

3.2 Let $x \in X$ be distinct from a. (Such a point exists since X contains at least two points.) The set $X - \{x\}$ contains the point a and is not equal to X, and so is not open. Thus $\{x\}$ is not the complement of an open set, and so is not closed. We deduce from Theorem 3.2 that $(X, \mathcal{T}_a)$ is not a Hausdorff space.

3.3 Let $(A, \mathcal{T}_A)$ be a subspace of a Hausdorff space $(X, \mathcal{T})$. If A is empty or contains just one point, there is nothing to prove. Otherwise, let $x, y \in A$ with $x \neq y$. Since $x, y \in X$ and $(X, \mathcal{T})$ is Hausdorff, there are disjoint sets $U, V \in \mathcal{T}$ with $x \in U$ and $y \in V$. Now $x \in U \cap A$, and $U \cap A \in \mathcal{T}_A$. Similarly, $y \in V \cap A$ and $V \cap A \in \mathcal{T}_A$. Since U and V are disjoint, it follows that $U \cap A$ and $V \cap A$ are disjoint. Thus $(A, \mathcal{T}_A)$ is a Hausdorff space.

3.4 Let $A \subseteq \mathbb{R}$ be bounded. Since $\mathbb{R}$ is unbounded, we can assume that $A \subset \mathbb{R}$. If $A = \varnothing$, then there is nothing to prove, so suppose $A \neq \varnothing$.

Suppose that A is bounded as defined in this unit. Then there exists $M > 0$ such that $|x - y| \leq M$ for all $x, y \in A$.

Hence, fixing $y \in A$ and using the Triangle Inequality, we find for any $x \in A$,
$$|x| = |x - y + y| \leq |x - y| + |y| \leq M + |y|.$$
Thus A is bounded according to the *Unit A1* definition.

Now suppose that there exists $M > 0$ such that $|x| \leq M$ for all $x \in A$. Then, by the Triangle Inequality,
$$|x - y| \leq |x| + |y| \leq M + M = 2M,$$
for all $x, y \in A$. Thus A is bounded as defined in this unit.

3.5 Let d_0 denote the discrete metric on $\mathbb{R}$. We know that $[0, 1]$ is closed (see Problem 1.5 of *Unit A4*) and since $d_0(x, y) \leq 1$ for all points $x, y \in \mathbb{R}$, by definition of the discrete metric, $[0, 1]$ is also bounded.

Every subset of $\mathbb{R}$ is open for the discrete topology — in particular, $\{x\}$ is an open set for each $x \in [0, 1]$. So $\{\{x\} : x \in [0, 1]\}$ is an open cover of $[0, 1]$ that has no finite subcover. Hence $[0, 1]$ is not compact.

4.1 We use the result of Theorem 4.1.

(a) $\{0\} \cup \{\frac{1}{n} : n \in \mathbb{N}\}$ is a closed set (see *Unit A4*, Worked problem 2.2). It is also bounded, since it is contained in $[-1, 1]$. It is therefore compact.

(b) Cl$(\mathbb{Q}) = \mathbb{R}$ (by Worked problem 2.1 of *Unit A4*). Hence using Theorem 2.6 of *Unit A4*, Cl$([0, 1] \cap \mathbb{Q})$ $=$ Cl$([0, 1]) \cap$ Cl$(\mathbb{Q}) = [0, 1] \cap \mathbb{R} = [0, 1] \neq [0, 1] \cap \mathbb{Q}$. Hence $[0, 1] \cap \mathbb{Q}$ is not closed, so it is not compact.

(c) $[0, 1] \cup [e, \pi] \cup \{100\}$ is a finite union of closed sets, and so is closed (see *Unit A4*, Theorem 1.4, Remark (ii)). It is also bounded, since it is contained in $[-100, 100]$. Hence it is compact.

(d) $[0, \infty)$ is not bounded and so is not compact.

4.2 We use the result of Theorem 4.2.

(a) $[0, 1] \times [1, 4]$ is a closed and bounded subset of $\mathbb{R}^2$, and so is compact.

(b) $\{(0, 0), (1, 0), (0, 1), (1, 1)\}$ is a closed and bounded subset of $\mathbb{R}^2$, and so is compact. (Alternatively, observe that $\{(0, 0), (1, 0), (0, 1), (1, 1)\}$ is a finite set, and so by the result of Problem 2.2 it is compact.)

(c) $\{(x, y) : x^2 + y^2 < 1\}$ is the unit *open* disc centred at the origin. It is not closed, and so it is not compact.

(d) $\{(x, y) : x^2 + y^2 = 1\}$ is a closed and bounded subset of $\mathbb{R}^2$, and so is compact.

4.3 Let p_1 and p_2 be the projection functions on $\mathbb{R}^2$. Then p_1 and p_2 are continuous and
$$f(x, y) = p_2((x, y)) \sin(\pi p_1(x, y))$$
$$\qquad\qquad + p_1((x, y)) \cos(\pi p_2(x, y)).$$

It is straightforward to use our list of basic continuous functions from $\mathbb{R}$ to $\mathbb{R}$ together with the Combination and Composition Rules for continuous functions from $\mathbb{R}^2$ to $\mathbb{R}$ to verify that
$$(x, y) \mapsto p_2((x, y)) \sin(\pi p_1(x, y))$$
and
$$(x, y) \mapsto p_1((x, y)) \cos(\pi p_2(x, y))$$
are continuous on $\mathbb{R}^2$ and so, by the Sum Rule, f is continuous on $\mathbb{R}^2$.

Hence, by the Restriction Rule, f is continuous on $\{(x, y) : 0 \leq x, y \leq 1\}$.

Also, $\{(x, y) : 0 \leq x, y \leq 1\}$ is closed and bounded. Hence, by Theorem 4.2, it is compact.

Thus both conditions of the General Extreme Value Theorem are satisfied, and we conclude that the function f attains a minimum value on $\{(x, y) : 0 \leq x, y \leq 1\}$.

Index

bounded set, 22

Cantor distance, 24
Cantor space, 19, 24
compact, 11
    closed subsets, 13
    Euclidean space, 26
    Hausdorff space, 21
    set, 11
    space, 11
compact subsets, 11, 13, 21
    intersection of, 23
compact surface, 29
compactness, 11
    topological invariance of, 13, 14
cover
    open, 5, 8

discrete topology, 18

Extreme Value Theorem, 9
    General, 27

finite subcover, 6, 14

good subset, 16

Hausdorff space, 18
    Cantor space, 19
    compact subset of, 21

continuous functions, 20
discrete topology, 18
metric space, 19
product of two such spaces, 20
subspace topology, 20
topological invariance of, 20
Heine–Borel–Lebesgue Theorem, 26
homeomorphism, 14, 20

indiscrete topology, 19

metric space, 19, 22
middle-third Cantor set, 25

open cover, 5, 8

separation axiom, 18
subcover
    finite, 6, 14
subset
    compact, 11, 21
    good, 16
surface
    compact, 29

Tikhonov's Theorem, 15, 27
topological space
    compact, 11
    Hausdorff, 18